Theory of Reality

By

John Paul Keyes

ISBN: 978-0-557-19345-5

Preface

This book is intended, as far as possible, to give an insight into a *Theory of Reality* to those readers who, from a general scientific and philosophical point of view, are interested in Reality - as opposed to Relativity - when it comes to better understanding the Universe that we live in.

Imagine a group of creatures living on this, or any planet, that do not have eyes or any concept of vision. These creatures have ears that are quite exceptional. Their sense of hearing, using sound waves, is as good as our sense of vision, using light waves. They instinctively understand the change in frequency of a sound wave caused by moving relative to the source of a sound. They call this the Doppler effect (or Doppler shift). They have excellent pitch and can determine how fast an object is moving by its Doppler shift.

I will call these creatures intelligent bats, to help you visualize their existence. They live in what I will call Batland.

Theory of Reality

Reality

"I can't take it anymore!" Johnbat suddenly blurts out to his brother.

"What?" Billbat asks.

Johnbat explains, "I have kept quiet about Albat's *Theory of Relativity* all my life. I've never believed in it, I've tried to ignore it, but I can't take it anymore."

"What do you mean?" Billbat asks again, puzzled.

"The idea that time is not absolute, that measurements change with velocity, that events that are simultaneous to one observer are not simultaneous to a different observer that is in motion - it's - it's - it's just ridiculous!" Johnbat rolled his ears then adjusted his ear horns. "It doesn't make any sense. I can't take it anymore. I have to say something. I have to tell someone. I just don't know how."

Johnbat was an accomplished mathematician and a computer scientist, but had little formal training in physics other than a few university courses. Billbat, on the other hand was a Ph.D. Chemist that had studied various wave properties and interference patterns, quite extensively. He was very familiar with Albat's theory of relativity. Both tried to keep up with their scientific backgrounds by reading science magazines and listening to educational shows on Public Radio.

The two brothers got along quite well with each other, despite the fact that they worked closely together.

"You know lots of scientists have tried to disprove relativity, but they always find that the formulas that Albat describe actually work when they are tested," Billbat offers.

"I know! I don't care! I still don't believe it and I just have to do something about it. I think that Albat made his formulas work, by changing the wrong variables. It all boils down to the speed of sound, or as they call it 'the propagation of sound waves'. Instead of logically figuring out that the speed of sound could not be fixed, he ended up changing time and distance for each observer, all to make it appear that sound traveled at the same speed regardless of any apparent motion by the observer. This makes the formulas work, even though, they are completely illogical. Albat should have shown that the speed of sound was different for each observer in motion, rather than time and distance being different."

Billbat adds, "That wasn't Albat's fault, you know."

"What?"

"The speed of sound. Albat did not decide that the speed of sound was fixed. This was given to him by others. Experimental evidence showed that the speed of sound was always the same regardless of your motion. They made the 'Law of Propagation of Sound' using these results. Albat basically said in his theory, that "if the Law of Propagation of Sound is true, then all this other stuff must also be true."

"Hmmm!" Johnbat frowns. "OK. I accept that. So, it wasn't Albat's fault after all. Then, I can only fault him for not figuring out that the 'Law of Propagation of Sound' was wrong. Well, I'm not sure that it is actually - wrong - I think that it's just being misinterpreted or misunderstood."

"Huh!" Billbat asks, "What do you mean?"

"Every time you try to measure the speed of sound, you always seem to get the same answer. It appears to be constant, no matter what your motion is. But, I believe this only happens because you are there to measure it. Like when you are traveling in a vehicle, you know that sound comes with you, in that vehicle? So, every time you try to measure its speed, it's always going to be the same."

Johnbat continues, "It's no different than tossing a ball or flying in a room, it does not matter how fast that room might be moving, you can still fly and talk normally. But, I don't think that anyone has tried to measure what is happening outside the room."

Johnbat sighs and starts thinking about what he wants to do. I know that I will sound like a fool to the world.

Theory of Reality

Who am I to question the great Albat. I have no credentials. It'd be like a hobo trying to tell me how to write an encryption routine. The hobo would have no basis for even understanding a small portion of how encryption even works. Yes, I have a mathematics degree. No, I do not have a physics degree. So, why do I feel that I have a leg to hang on? Well! I can do thought experiments better than anyone I know. I am a logic expert. And, I can tell you without hesitation that relativity is not logical. Even Albat knew that.

I have gone back through Albat's theory of relativity, very carefully. I felt that something was wrong 30 years ago, when I first studied it. I didn't say anything then because I was afraid of being ridiculed. Now that I am arguably smarter and more experienced, or at least bolder than when I was younger, I figured that I should understand and appreciate his theory better. Well, I understand it all right! But, I still cannot agree with his conclusions. I believe that I've found problems with his logic. At each phase of his quest to define his theory, Albat has to make logical choices of direction to explain experimental or expected results. I think that he makes the wrong choices in a number of cases.

I submit that time is absolute. It is not relative to a reference-body, as Albat asserts. That is just ludicrous. Yet, Albat has built all his formulas and theories around the idea that time is not absolute. That just by moving, you cause time to slow down. In fact, he specifically claims "there is no meaning in a statement of the time of an event." When confronted with the apparent incompatibility of the law of Propagation of Sound with the Principles of Relativity, Albat makes the decision that time and distance must be different for each reference body.

Before the advent of the theory of relativity it had always been assumed in physics that time was absolute. That it was independent of the state of motion of the body of reference. But, Albat discards this assumption of time being absolute, in favor of his definition of simultaneity and the law of the propagation of sound. This, I believe, is where he makes his first fatal mistake. Logic tells me that either his definition of simultaneity or the law of propagation of sound is flawed. One or *both* must be wrong. Because in my logical world -

Time is absolute.

Background

Let's start with a bit of background. Albat's theory of relativity starts off by describing Euclidean geometry. This leads into a definition of a Galileian coordinate system based on a rigid body of reference. The *principle of relativity* states that the laws of physics have the same form in all bodies of reference. I don't have a problem with this, in fact no empirical data has ever been found which is contradictory to this principle.

Many experiments have shown that if you are in a closed room, for example in a hold on a ship at sea, there is no way for you to know if that room is moving or not. All the laws of nature work exactly the same within that room, as within any other reference-body. When you bounce a ball, or you fly - everything will always move in the exact same manner, regardless of any direction or speed that the room itself might be moving.

In order to describe his theory of relativity Albat uses a lot of thought experiments. In some of Albat's examples, he describes two different bodies of reference; a railway platform with embankment is one body of reference and a railway carriage that passes by the railway platform as another body of reference.

Albat first attacks classical mechanics, when he claims that "the theorem of addition of velocities, cannot be maintained." Let's review that law for a moment, using his example:

Let us suppose that a bat is located on a railway carriage that is traveling along the rails with a constant velocity v, and that the bat flies the length of the carriage in the direction of travel with a velocity w.

How quickly, or with what velocity W does the bat advance relative to the embankment during the process? The only possible answer seems to result from the following consideration: If the bat were to land for a second, he would advance relative to the embankment through a distance v equal numerically to the velocity of the carriage. As a consequence of his flying, however, he traverses an additional distance w relative to the carriage, and hence also relative to the embankment, in this second, the distance w being numerically equal to the velocity with which he is flying. Thus in total he covers the distance $W = v + w$ relative to the embankment in that second considered.

Albat insists that this addition of velocities does not hold in reality. He goes on to demonstrate this with his formulas. I submit that the addition of velocities does hold in reality. Albat changes time and distance for any object in motion. I do not believe that we have to take this illogical path. I intend to show that no dilation of time or length contraction actually occurs. I believe that the observations that appear to demonstrate these phenomena are flawed because of a misunderstanding of the law of propagation of sound. Also, I believe that the use of sound waves in our observations may limit our ability to hear what is really happening.

I like to imagine that if we had a better method of observing events with tools that work better than listening to sound waves, that we would witness reality, rather than relativity. After all, by placing an observer perpendicular to the two platforms, it is possible to hear the actual events more clearly. Things do not sound the same for the original observers because of the way sound waves can misrepresent their results.

The Law of Propagation of Sound

There is hardly a simpler law in physics than that to which sound is propagated in dry air. Every school pup knows, or believes they know, that this propagation takes place in straight lines with a velocity that is roughly *c=343 meters per second.* By means of careful observations it has been determined that the velocity of propagation of sound cannot depend on the velocity of motion of the body emitting the sound. Many experiments have shown this to be the case.

For example, two bats flying in opposite directions both squeak just as they pass close by each other. An observer will hear both squeaks, at the same time regardless of the motion of the bats or the location of the observer. Their squeaks will not travel any slower or faster because of the speed of their flight. Their squeaks always move at the same speed, the speed of sound. So, we hear them at the same time.

This simple law of the constancy of the velocity of sound has plunged physicists into the greatest intellectual difficulties. Let us consider how these difficulties arise.

Let us refer the process of the propagation of sound to a rigid reference-body (coordinate system). We will choose Albat's embankment. We shall imagine the air above it to be still and dry. If a sound wave was sent along the embankment, we hear from above that the tip of the sound wave will be transmitted with the velocity c relative to the embankment. Now let us suppose that our railway carriage is again traveling along the railway with a velocity v, and that the direction is the same as that of the sound wave, but with a slower velocity. Let us inquire about the velocity of the propagation of the sound wave to the carriage. It is obvious that we can here apply the consideration of the previous section, since the sound wave plays the part of the bat flying along relative to the carriage. The velocity W of the bat relative to the embankment is here replaced by the velocity of sound relative to the embankment. (w) is the required velocity of the sound with the respect to the carriage, and we have

$$w = c - v$$

The velocity of the propagation of a sound wave relative to the carriage thus comes out smaller than c.

But this result appears to come into conflict with the principle of relativity. For, the law of the transmission of sound must, according to the principle of relativity, be the same for the railway carriage as reference-body as when the rails are the body of reference. But, from our above observation, this would appear impossible. If every sound wave is propagated relative to the embankment with the velocity c for this reason it would appear that another law of propagation of sound must hold with respect to the carriage - a result contradictory to the principle of relativity.

In view of this dilemma, it appeared to Albat that we must abandon either the principle of relativity or the simple law of the propagation of sound. At this juncture Albat throws the theory of relativity into the cave. By altering time and distance for the object in motion, he was able to come up with formulas that make both the principle of relativity and the law of propagation of sound work together.

That was a total cop-out, in my opinion. Instead of making the hard decision, to point out that the law of propagation of sound must somehow be flawed, Albat decides to change time. That was the beginning of the end. Nothing after that is reasonable. When you get the fundamental assumptions wrong, everything else is worthless. We use a phrase in the computer world; 'garbage in - garbage out.'

I submit that by better understanding reality we should

be able to make a new law of propagation of sound that works with the principle of relativity without altering time or distance for either body of reference.

Why couldn't Albat have realized that the principle of relativity was only supposed to work within a closed room or system? What if the railway carriage was not a closed system? The sound waves were moving outside the carriage, not inside it. So, the principle of relativity should not play a part.

Grubland

"I've decided to write a book," Johnbat tells his brother.

"You mean about how you think Albat's theories are wrong?" Billbat asks.

"Yah, except that I've got a great idea!"

"What?"

"Up until now, I couldn't figure out how I could present my ideas. You know that I can't just write a physics book or present a physics paper. I have no credentials or credibility and it'd open me up to ridicule by the scientific community. I haven't really studied this all my life, like some of them have. And, I know that Albat's formulas actually work. As you know, I just don't believe in the reasons behind the formulas, especially his use of time dilation."

"So, what's your idea?" Billbat asks.

"I think I'll write the book about an imaginary creature. This creature won't be able to hear, like we can. His entire life would be centered around smelling, let's say.

So, all his laws and formulas will be based on the speed of smell. But, his plight can be extrapolated into our own world. By showing that his views of reality were limited because they were based on the speed of smell, I intend to demonstrate that we may have limitations as well."

"That sounds like a neat idea."

"And, it has some advantages. Not many bats have a firm grasp on sound waves and exactly how they are propagated through various materials. It should be much easier for everyone to grasp how smell particles work rather than sound waves, even though the analogy is not perfect. I expect that they will be able to hear through the analogy to understand what is really being presented. And, who knows, maybe it will inspire some real physicist to be intrigued by the ideas that I present and take the ideas further, into a new theory of reality."

"Hmmmm! Not bad," says Billbat. "So, how would it work?"

"The book would go something like this;"

Imagine, if you will, a group of creatures living on this, or any hill, that do not have ears or any concept of sound. These creatures have noses that are quite exceptional. Their sense of smell is as good as our sense of hearing. In fact they can determine exactly how far an object is away from them and how fast it is moving. They knew the gaseous diffusion rates for the speed of smell and they can detect how much the smell particles have degraded since they left the source, to determine its distance and speed.

I will call these creatures intelligent grubs, to help you conceptualize their existence. They live in what I will call Grubland.

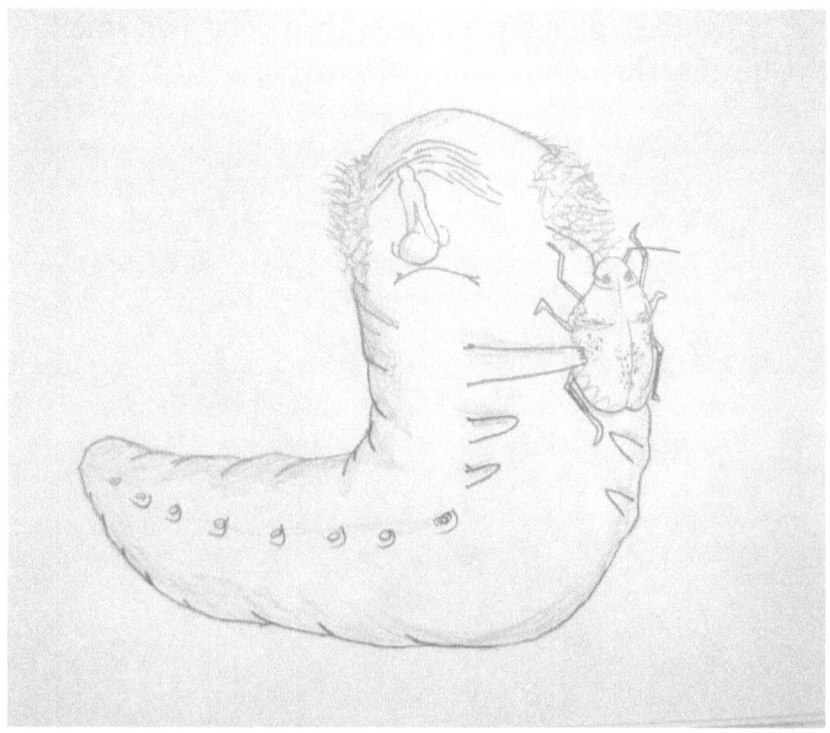

Theory of Reality

These grubs will necessarily define their theories based on the best method of observation that they know, namely the speed of smell. Every school grub knew that the speed of smell from a standard stinkbug through a dry tunnel was roughly 30mm per second.

One of their smartest was called Algrub. Algrub's relativity theories stated that nothing could exceed the speed of smell. All underground tests had confirmed his theories. It was obvious to everyone that you could not crawl through a tunnel faster than your own smell. So, his theories could not be disproved.

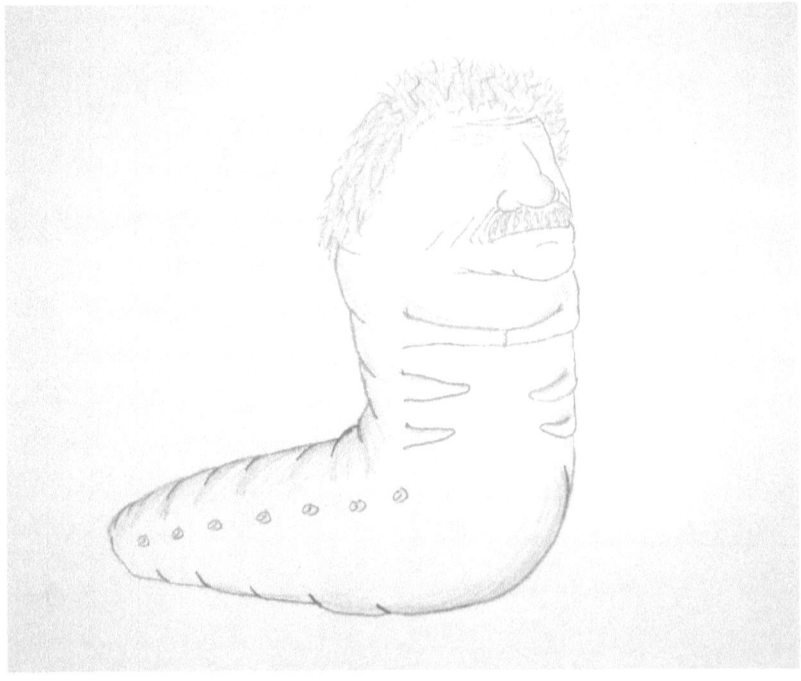

Theory of Reality

Can you imagine the shock that Algrub might experience if he happened to surface one day while you or I were flying above him? As you know, we fly much faster than the speed of smell, on a regular basis. I've been thinking about what he might witness.

If I were to swoop down near enough to Algrub at let's say 90mm/second, 3 times faster than the speed of smell, I believe that he would witness a very strange event. I figure that he would detect the spontaneous appearance of a bat and what I will call an anti-bat flying off in opposite directions. The bat would proceed in the direction that I was actually going, the anti-bat would appear to leave in the direction that I came from.

Because of the inherent limitations with respect to the speed of smell, Algrub would be quite confused by what he witnesses. He would try to determine what had happened. After 1 second I would have traveled 90mm away. My smell would then take another 3 seconds to come back to Algrub. He would assume that I was traveling - let's say 90mm/ 4 seconds or 22.5 mm/second. The anti-bat would be even weirder. My smell, from 90mm away, or 1 second before I swooped past him, would reach him just 2 seconds later. He would then calculate 90mm/ 2 seconds or 45mm/ second speed. This would be faster than the speed of smell.

Billbat appeared confused. He fluttered around the room drawing figures in the air. "What's wrong?" I asked.

"I think you've made a mistake, in the calculations." Billbat expressed.

"What do you mean? 90mm/2 seconds = 45mm/second. What's wrong with that?"

"You said that Algrub was intelligent. Right?"

"Yah! So?"

So Billbat explained.

When you passed Algrub he would first smell you at second 0. When you were 90mm away, 1 second later, it would take another 3 seconds more for the smell to come back to Algrub. But, Algrub would know that this was the case. His intelligent brain would deduce that you must have traveled 90mm in just 1 second. But this would violate Algrub's accepted theory of relativity.

Now for the so called anti-bat: It would appear to be leaving in the direction that you actually came from. Algrub would detect the anti-bat at 90mm away, in just 2 seconds. This would be quite confusing. When he tried to mentally subtract the 3 seconds that he knows that the smell must have traveled to reach him, nothing would make any sense. He might conclude that the smell particles must have been traveling faster than the speed of smell. But this would violate the law of propagation of smell. So, he might decide that the rate of smell particle decay or time was wrong for the fast moving object. He would not be able to figure out that he had witnessed an object that was actually moving faster than the speed of smell.

"Hmmm! You're right," Johnbat admitted to his brother. "But, I can fix that. I'll just turn around, swoop back to pick up Algrub and eat him before he smells me coming."

"End of confusion."

Simultaneity

Let's go through a thought experiment about
simultaneity. The famous one that Albat used will do
just fine. Albat makes an assertion that two events (like
two explosions) occur simultaneously. The question
arises, after some consideration: how can one determine
by observation that the two events actually took place
simultaneously or not?

The concept of 'simultaneous' does not exist for a
physicist until he has a way of testing it in an actual
case. Albat decides to define 'simultaneity' such that the
definition itself supplies us with the method in which
we can test that two events occur simultaneously.

Albat thinks about it for some time, he then makes the
following suggestion on how to test simultaneity. By
measuring along the rails, the connection AB should be
measured and an observer placed at the mid-point M
between points A and B.

This observer listens for the two explosions that will
occur at the places A and B. If the observer at M hears
the two explosions at the same time, then he concludes

that they were simultaneous.

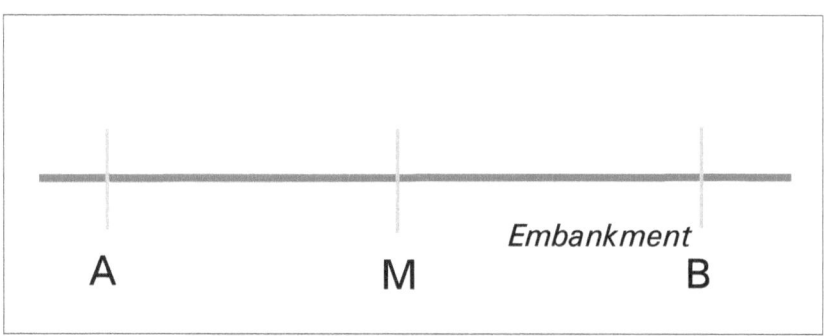

Albat realizes that this definition of 'simultaneity' would only be right if he knew that sound traveled along the length A->M with the same velocity as along the length B->M. Albat also realizes that this supposition would only be possible if we already had a means of measuring time. He decides that he is moving in a logical circle.

He then declares that his definition of 'simultaneity' must be correct, because it does not assume anything about the speed of sound. In his definition there was only one demand, that in every real case it supplies us with a method to determine 'simultaneity'. He believes that his definition satisfies this demand. He says, and I quote "that sound requires the same time to traverse the path A->M as for path B->M is in reality neither a supposition nor a hypothesis about the physical nature of sound, but a stipulation which I can make of my own free will in order to arrive at a definition of simultaneity."

This is a huge and fatal mistake in logic. He has made up, off the top of his head, a definition of 'simultaneity' that is flawed, because it requires or stipulates that the constancy of the speed of sound is correct. It is quite easy to disprove this stipulation. His own thought experiment showed that. Here it is:

We suppose a long flatbed train is traveling along the rails with a constant velocity v and in the direction indicated in the drawing that follows. Observers traveling on this train will use the train as a rigid reference-body (coordinate system); they regard all events in reference to the train. Every event which takes place along the line takes place at a particular point of the train. Also the definition of simultaneity can be given relative to the train in exactly the same way as with respect to the embankment. As a natural consequence, however, the following question arises:

Are two events (example: the two explosions at A and B) which have been defined as simultaneous with reference to the railway embankment also simultaneous relative to the train? Albat attempts to show directly that the answer must be NO. I intend to prove that his definition of simultaneity was flawed, that the speed of sound was not constant, and in fact that any two events that are actually simultaneous are simultaneous in all frames of reference.

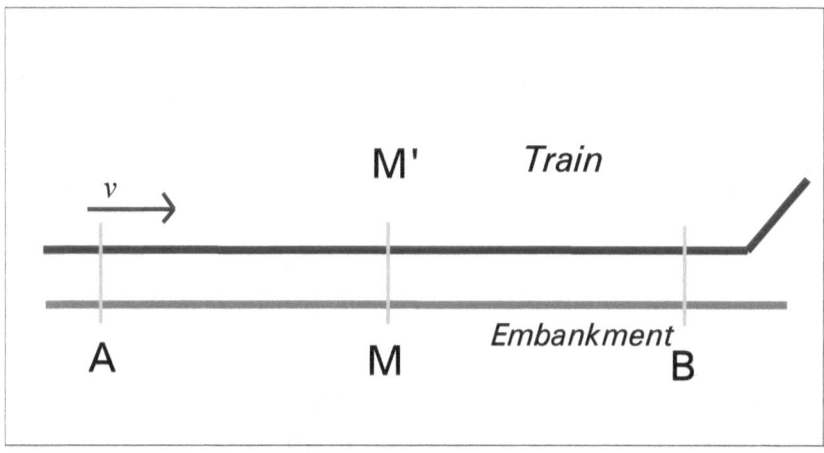

Remember, Albat defines simultaneity as an observer hearing two explosions (that were equidistant from the observer) at the same time.

When we say that the explosions at A and B are simultaneous with respect to the embankment, we mean: the sound waves emitted at the places A and B, where the explosions occur, meet each other at the mid-point M of the length A->B of the embankment. But the events A and B also correspond to positions A and B on the train. Let M' be at the mid-point of the distance A->B on the traveling train. Just when the explosions occur, this point M' naturally coincides with the point M, but it moves towards the right in the diagram with the velocity v of the train. If an observer sitting in the position M' on the train did not possess this velocity, then he would remain at M, and the sound waves emitted by the explosions would reach him simultaneously.

Albat continues, now in reality (considered with reference to the railway embankment) he is hastening towards the sound wave coming from B, whilst he is riding on ahead of the sound wave coming from A. Hence the observer will hear the sound emitted from B earlier than he will hear that emitted from A. Observers who take the railway train as their reference-body would therefore come to the conclusion that the explosion at B must have taken place earlier than the explosion at A. Albat therefore arrived at his first important mistaken result that:

Events which are simultaneous with reference to the embankment are not simultaneous with respect to the train, and vice versa. This is Albat's (relativity of simultaneity). He concludes that every reference-body (coordinate system) has its own particular time; and therefore unless we are told the reference-body to which the statement of time refers, he suggests that there is no meaning in a statement of the time of an event.

Albat came to this illogical conclusion because of his

stipulation that sound requires the same time to traverse a path A->M as for the path B->M. He then used that stipulation (which I intend to prove false) to arrive at his definition of simultaneity. He also tries to use the railway train as a reference body, without proving that it actually could be used as such. This is how he allows himself to deviate from reality; he starts with an invalid definition, adds a reference body of questionable validity and forces his math to explain it all.

My conclusions are very much different from Albat's. The actual and true definition of simultaneous can be taken from our own Batster's Dictionary. It is supposed to mean 'events that occurred at the *same time*.' I submit that simultaneity cannot be defined by simply saying that the sounds from two equidistant events were heard at the same time. This, in my mind, is no proof that those events did or did not occur at the same time.

Remember, I am trying to demonstrate that sound does not have a constancy of speed, except within a given body of reference. Albat's invalid definition of simultaneity does not take into consideration a number of important possibilities. That sound waves will travel through any particular medium at the same rate of speed. But, that the speed of sound is affected by the movement and density of the medium that it is traveling through. That if you are advancing towards a source of sound waves, that you will reach those waves earlier than if you were stationary or traveling away from the source. Therefore, the rate of speed of sound will depend on whether or not the observer was moving within a valid body of reference.

Albat incorrectly chose the flatbed train as a reference body. Whereas, I believe that the platform itself was the only valid reference body, in that case. Remember that the principle of relativity was only supposed to work within a closed body of reference. The flatbed train was open to the world so that anyone could easily tell it was moving relative to the sound waves.

Closed versus Open Systems

The most important factor, that I believe Albat seemed to misunderstand, was the concept of closed versus open systems. I have no argument with the laws of physics which state that they are the same in any uniformly moving reference body. All tests to confirm this were done in closed rooms. The problem has always been knowing what body of reference that the observer was really located in. If you are in a vehicle traveling at any velocity, the speed of sound will always be the same inside that vehicle. However, the speed of sound will not be the same outside that vehicle. If you stick your head out of the door of a moving vehicle, you are no longer inside the vehicle's body of reference. Figuring out what medium a sound wave is actually traveling through is the key to knowing how fast it may be moving, at any particular time or place.

Let us go through another thought experiment to demonstrate this: Let us suppose that the train is not open to the environment, but is closed, or at least mostly closed off from the embankment. Let's say that it has a top and sides that run the complete length of the train like a tube, with closed ends. This will allow the train to drag its environment along with it. It will therefore become a closed system and a valid body of reference. At this point, the principle of relativity will work inside the train. However, for our experiment, we will give the train doors that can be opened that face the embankment.

When we say that explosions occur at A and B simultaneously, we will mean that they occurred at exactly the same time. And, for this experiment, we will say that the sound will reach the mid-point M of length A->B of the embankment at the same time.

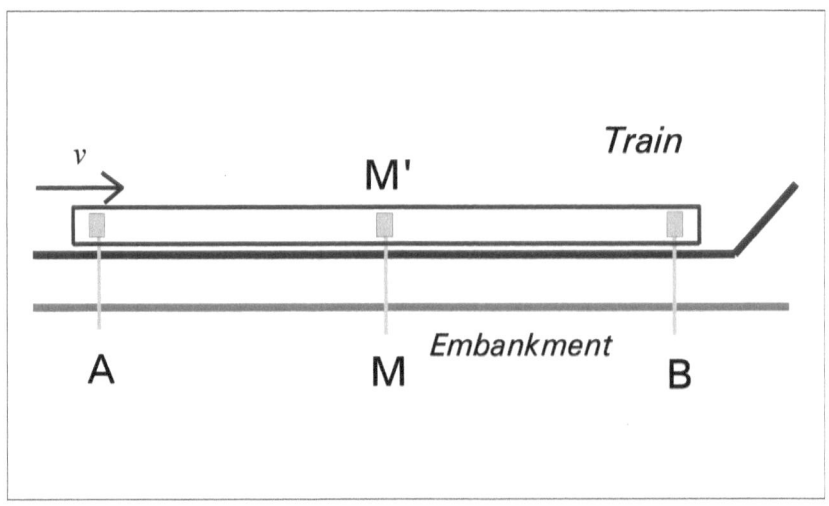

If the sound waves were to travel through open doors at

positions A and B of the train, those sound waves that entered the closed train would necessarily change their speed relative to the velocity of the train itself. The observer at M' inside the train would therefore hear the sounds from A and B at the same time and would conclude that events A and B occurred simultaneously, according to Albat's original definition of simultaneity. This conclusion would completely agree with the observer at position M on the embankment even though the observer at M' had been moving relative to M. We could now conclude that simultaneity is not relative to the reference-body, but is absolute just as time is absolute.

If we went further and opened a third door right at location M', the observer at this location would hear four different explosions. First he would hear the explosion coming from B, that traveled outside the train and passed through the open door at M'. He would then hear both explosions simultaneously coming from A and B that had passed through the open doors at positions A and B and traveled inside the train. And finally, he would hear the explosion coming from A that traveled outside the train and then passes through the open door at M'.

Using the difference in time from each of those events, it could even be possible for the observer at M' to calculate his velocity v relative to the embankment, where the original explosions had taken place.

When the law of propagation of sound claimed that sound always traveled at the same speed, regardless of the motion of the observer, this was determined by observations in a closed environment. It would have been so easy to show that if you are moving within a closed system - that sound waves do not come at you at the same speed as it does to stationary observers within that same system. The original open train example demonstrates this quite convincingly.

Practical Joke

"A strange thought has come to me," Johnbat says. "I wonder if I should include it in the book."

"What's that?" Billbat asks.

"I kinda wondered if Albat was joking," Johnbat paused.

"Joking?" Billbat asks frowning.

"Well, I've been thinking that Albat might have been joking when he wrote the theory of relativity. He must have known that the only way that sound could hit everyone at exactly the same speed even if they were all moving in different directions, would be if sound had magical properties. So, he must have known that the law of propagation of sound was somehow flawed. It didn't make sense." Johnbat explains.

Billbat just hung there, puzzled.

"OK! Try this out for size," Johnbat explains. "What if he wrote the theory of relativity as a joke to point out that the law of propagation of sound was no good and stupid. But, he was so brilliant, that he made the math work too well. Instead of his colleagues laughing about it and realizing their mistake, they all took him seriously. In fact they hung him on a pedestal so fast that Albat couldn't take it back. And, to this day, no one has figured it out, because his formulas were that good."

There was a long pause, as Billbat puzzled over what I had just said. Then ... "Nah!" Billbat chuckled shaking his head. "I don't think you should say that. It might be taken as disrespectful."

"I suppose you're right." Johnbat was visibly disappointed. "I better not include it. But the idea is so funny. I don't know how I can bite my tongue."

"What about the Math?" Billbat asks. "Are you going to include any math in the book?"

"I don't really want to include any real math in the book," Johnbat answers. "Remember, it isn't going to be a physics book, and I believe that most readers would find the math boring. But I suppose I should say something about the math. After all, relativity is defined by mathematics."

"It is obvious that the math that Albat included in his theory actually works when you test it. Albat was a super-genius after all. However, I believe that it works because his math specifically targets the observational limitations that fixing the speed of sound (c) causes. If he had put the same effort into allowing the speed of sound to change with each observer, instead of forcing time and distance to change with motion, the math would also work. The new math would be completely consistent with classical mechanics," Johnbat concluded.

"Assuming that you were flying so high that you could not fix your position with the ground below, Albat could not figure out where your valid body of reference would be. His theory of relativity eliminated this mystery, by declaring it irrelevant. Each observer would become their own body of reference, and would therefore consider themselves not moving. He decided that sound waves would always hit an observer at the same speed regardless of their possible motion relative to the sound waves. In other words, he made the decision that if you couldn't figure out what your valid body of reference was, then you declared yourself the body of reference. This, in essence, meant that you were not moving. But, neither was the bat that just flew past you. You would think that he was moving, but he would think that you were moving. In order for this to work, something had to give. Albat chose time and distance as the variables."

"Hear what has happened because of the theory of relativity. A whole colony of nerdbats have been produced that spend their entire existence making up stories of time travel, folded space and worm holes. They make up all kinds of paradoxes and contradictions as the result. All of this because Albat did not recognize that the law of propagation of sound was invalid."

Johnbat continued, "Albat's math works by ignoring reality. Relativity works because it explains your observations without knowing where your true body of reference was. My complaint is that most of us use relativity without realizing that the bending of time and space is an observational illusion. It is not reality."

Time Dilation

Batwell, a great physicist, demonstrated that sound traveled at a constant speed (c) in still air. This had a major consequence. Batwell's equations gave the speed of sound to be 343 meters per second. Therefore, demanding that the laws of physics were the same in all inertial frames implied that the speed of any sound wave in any inertial frame will always give 343 meters per second. Remember, the principle of relativity says that when you are in a closed room, you cannot tell if that room is moving - all the laws of physics will be the same within that room.

Let us suppose that you were flying way up high, so high that you could not fix your location with the land below; Batstein, Michelbat, Morbat, Batentz - no physicist could figure out where your frame of reference would be. They concluded that there would be no fixed frame of reference that everything could be measured against, when you were that high. They concluded that sound would always travel at the same speed, regardless of your potential motion. This incorrect conclusion had some dramatic and surprising consequences.

Albat was faced with the apparent incompatibility between the law of propagation of sound and the principle of relativity. At this point, it should have been obvious that the law of propagation of sound had to be tweaked, something was wrong with it. Instead of that, he decides to drop two of what he calls unjustifiable hypothesis from classical mechanics; as follows:

1) The time-interval (time) between two events is independent of the condition of motion of the body of reference.

2) The space-interval (distance) between two points of a rigid body is independent of the condition of motion of the body of reference.

Albat makes the illogical decision to drop these hypothesis, in favor of the unproven law of propagation of sound. In my opinion the experiments that led to the law of propagation of sound were being unjustifiably generalized. Although the speed of sound in all experiments up to that time had shown that it was the same, I propose that the speed of sound could be different in other circumstances. I would propose the law of propagation of sound could be reworded as such: Sound always travels at the same speed *within any body of reference*, regardless of the condition of motion of that body of reference.

Let us go back to the railway platform and the railway carriage for a quick example. You must indicate which body of reference any sound wave is traveling within, in order to fix its speed. Sound will always travel the same speed (c) within the closed railway carriage. But, just outside the railway carriage walls, sound waves will be traveling in reference to the embankment. This means that if you remove the carriage walls, the sound waves would be traveling at c-v or c+v, consistent with classical mechanics.

Once you recognize that sound waves change their velocity to match the frame that it is traveling within, everything starts to make sense. The confusion was that each time you try to test the speed of sound, you test it in either a closed room or using closed equipment that drags the sound with it. Just by being there you change its speed and therefore you change the results. A valid test of the speed of sound would have to be done in such a way that the test does not change the speed to match the observer or the equipment. The only way to do that is to know exactly what makes it follow you. Running a flatbed train with an observer on it demonstrates this point quite easily. It is apparent that sound is not traveling the same speed both for the observer on the platform and for the observer on the open train. If you are moving relative to another fixed frame, sound will reach you at a different rate of speed.

Supersonic Travel

Johnbat and Billbat sat on the top of a hill, well above the home caves below. They were both pondering and generally thinking about what they had been discussing over the past week.

"Hey Billbat?"

"Huh?"

"Remember the intelligent grubs that we talked about, earlier?"

"Yah."

"I've been thinking a lot about what poor old Algrub would think of super-aromatic travel. You know - traveling faster than the speed of smell. In his mind, he felt that it was impossible. Especially in the conditions where he lives - you know - underground and all that. It makes me think that it might be possible for us to actually move faster than the speed of sound, someday."

"If something could exceed the speed of sound, don't you think that a supersonic event would have been

43

observed by someone, by now? And believe me, plenty of scientists have tried to find evidence of this." Billbat replied with an even temperament. Billbat never ridiculed his brother's ideas. He would just point out what was generally accepted by respected scientists.

"Well, just what if we could? How would we do it? Obviously we could not use any of the propulsion systems that we now possess. We would have to invent something new," Johnbat explains, in his science fiction type imagination.

"Let's talk about what we have today; We flap our

wings to move through the air. No matter how fast you flap a wing, we are not going to get even close to the speed of sound, even diving straight down from a great height. What if we could make something like a Sound Drive. After all, sound is the fastest thing we know of. It would send out sound waves that push a vehicle faster and faster."

"The problem with this is that the amount of sound being transmitted would have to increase exponentially to get even close to the speed of sound. But, you would never be able to actually reach the speed of sound using this method, no matter how much energy you pumped into the sound drive. The amount of acceleration that you would get from a Sound Drive would get smaller and smaller, the faster you went."

"Even if you could get all the way up to speed, what do you think might happen?"

"I don't know. I guess that there would be a huge Doppler effect, as you went faster and faster. Eventually all the sound waves would be piled upon themselves. I would imagine that this sound wave would get so immense that it would rip you apart. Who knows what those piled up waves would even sound like to any observer that you passed. Nope! Albat must be right. There is no way that you could travel faster than the speed of sound."

They both turned to smell the blossoms, while they sat listening to the world below. Johnbat seemed disappointed in their conclusions. But, just then something incredible happened! They heard some sort of an explosion, just off the face of the cliff, in front of them.

Then they both heard the same thing; two enormous hard surfaced objects just appeared out of nowhere, directly in front of them. They produced an incredibly load noise, like a tornado at a caves entrance. You could hear them speed off in opposite directions, like an object and an anti-object, very fast... Too fast.

Johnbat and Billbat turned to each other. Their mouths opened in amazement. They just sat there, too startled to even squeak.

Epilogue

We don't know everything. What we should know, however, is that light is not magic. It cannot have a constant speed regardless of the observer's motion. Einstein developed special relativity followed by general relativity based on the premise that light would always hit an observer (in space) at the same speed, even if the observer was moving. Let's look at what led Einstein to the belief in the Constancy of the Speed of Light.

Luminiferous aether is the theoretical medium that propagates light. In 1887 the Michelson-Morley experiment was designed to try to measure the aether wind, using an interferometer. It failed to detect the aether. But the experiment failed because it was done on Earth within its atmosphere. In 1851 the Fizeau experiment had already shown that light speed was affected by moving fluids. This meant that light should be affected by the movement of Earth's atmosphere. When the interferometer could not detect the aether wind, it was falsely concluded that there was no aether.

I believe that the elusive luminiferous aether does exist, in the universe. It cannot be an accident that light travels at the same speed from all directions, here on Earth. Something must be causing this effect. I believe that gravity is the aether that light travels on. Gravity is the one thing that is everywhere, and blocked by nothing.

As light travels throughout the universe, it may speed up or slow down to match each gravitational field that it enters or passes. For this reason, we would always witness the speed of light to be the same, within our own system. We must be careful not to assume that light travels the same speed within other stronger or weaker gravitational fields. And, we cannot know how fast light travels between gravitational fields, such as between star systems or between galaxies. It is possible that when light travels for a very long time through very weak gravitational fields between galaxies, that it might degrade little by little causing a red-shift. If this were the case, the universe might not even be expanding. The Big Bang may never have occurred.

We can see evidence that gravity affects the speed of light when you view a star that is behind a strong gravitational field, like our sun or a black hole. When light passes through a glass lens, it bends the light by slowing it down. It has been shown that light that passes near a strong gravitational field will bend. This is referred to as a gravitational lens. Space does not have to warp near a gravitational field, the light would simply slow down, which in turn would cause a bending of the light.

Within the Event Horizon of a Black Hole, space would not have to fold upon itself. The gravitational field might be so strong that the light cannot escape. Light might either slow down to zero, or it may bend in a circle around the Black Hole.

Assuming that gravity is proven to be the elusive aether, this would completely nullify much of Einstein's concepts that are not logical. Einstein's work was based on the constancy of light. This misunderstood concept caused all of the problems. He could not figure out what fixed body of reference could be chosen when you were in so called empty space. By deciding that there was no aether, no fixed frame of reference; he had to change time for each reference-body in order to make light travel at the same speed for all observers with any apparent motion. I believe that if we could stick our heads out of our moving vehicles, we would witness light traveling at different speeds, depending on our motion.

When you have two different observers, that are in motion relative to each other, both will not have to think that time slows for the other. In fact, neither will experience a time dilation. Clocks and chemical processes may run at different rates, due to their speed and/or their location within a gravitational field, but time itself remains constant throughout the universe. When two observers stop moving relative to each other, time is still the same, regardless of their respective clocks.